Connected Mathematics

Prime Time

Factors and Multiples

Student Edition

Glenda Lappan
James T. Fey
William M. Fitzgerald
Susan N. Friel
Elizabeth Difanis Phillips

Developed at Michigan State University

DALE SEYMOUR PUBLICATIONS®

The Connected Mathematics Project was developed at Michigan State University with the support of National Science Foundation Grant No. MDR 9150217.

This project was supported, in part, by the
National Science Foundation
Opinions expressed are those of the authors and not necessarily those of the Foundation

The Michigan State University authors and administration have agreed that all MSU royalties arising from this publication will be devoted to purposes supported by the Department of Mathematics and the MSU Mathematics Education Enrichment Fund.

This book is published by Dale Seymour Publications®, an imprint of the Alternative Publishing Group of Addison-Wesley Publishing Company.

Managing Editor: Catherine Anderson
Project Editor: Stacey Miceli
Production/Manufacturing Director: Janet Yearian
Production/Manufacturing Coordinator: Claire Flaherty
Design Manager: John F. Kelly
Photo Editor: Roberta Spieckerman
Design: PCI, San Antonio, TX
Composition: London Road Design, Palo Alto, CA
Illustrations: Pauline Phung, Margaret Copeland, Mitchell Rose, and Ray Godfrey
Cover: Ray Godfrey

Photo Acknowledgements: 11 © Brian Baer/UPI/Bettmann; 13 © Ray Massey/Tony Stone Images; 24 © A. Rezny/The Image Works; 26 © Akos Szilvasi/Stock, Boston; 37 © The Bettmann Archive; 41 © John Kelly/Tony Stone Images; 55 © Mike and Carol Werner/Comstock

Copyright © 1996 by Michigan State University, Glenda Lappan, James T. Fey, William M. Fitzgerald, Susan N. Friel, and Elizabeth D. Phillips. All rights reserved. No part of this publication may be reproduced, stored in a retrieval system, or transmitted, in any form or by any means, electronic, mechanical, photocopying, recording, or otherwise, without prior written permission of the authors. Printed in the United States of America.

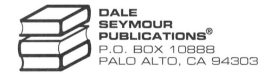

DALE SEYMOUR PUBLICATIONS®
P.O. BOX 10888
PALO ALTO, CA 94303

Order number 21441
ISBN 1-57232-146-6

1 2 3 4 5 6 7 8 9 10-BA-99 98 97 96 95

The Connected Mathematics Project Staff

Project Directors

James T. Fey
University of Maryland

William M. Fitzgerald
Michigan State University

Susan N. Friel
University of North Carolina at Chapel Hill

Glenda Lappan
Michigan State University

Elizabeth Difanis Phillips
Michigan State University

Project Manager

Kathy Burgis
Michigan State University

Technical Coordinator

Judith Martus Miller
Michigan State University

Curriculum Development Consultants

David Ben-Chaim
Weizmann Institute

Alex Friedlander
Weizmann Institute

Eleanor Geiger
University of Maryland

Jane Mitchell
University of North Carolina at Chapel Hill

Anthony D. Rickard
Alma College

Collaborating Teachers/Writers

Mary K. Bouck
Portland, Michigan

Jaqueline Stewart
Okemos, Michigan

Graduate Assistants

Scott J. Baldridge
Michigan State University

Angie S. Eshelman
Michigan State University

M. Faaiz Gierdien
Michigan State University

Jane M. Keiser
Indiana University

Angela S. Krebs
Michigan State University

James M. Larson
Michigan State University

Ronald Preston
Indiana University

Tat Ming Sze
Michigan State University

Sarah Theule-Lubienski
Michigan State University

Jeffrey J. Wanko
Michigan State University

Evaluation Team

Diane V. Lambdin
Indiana University

Sandra K. Wilcox
Michigan State University

Judith S. Zawojewski
National-Louis University

Teacher/Assessment Team:

Kathy Booth
Waverly, Michigan

Anita Clark
Marshall, Michigan

Theodore Gardella
Bloomfield Hills, Michigan

Yvonne Grant
Portland, Michigan

Linda R. Lobue
Vista, California

Suzanne McGrath
Chula Vista, California

Nancy McIntyre
Troy, Michigan

Linda Walker
Tallahassee, Florida

Software Developer

Richard Burgis
East Lansing, Michigan

Development Center Directors:

Nicholas Branca
San Diego State University

Dianne Briars
Pittsburgh Public Schools

Frances R. Curcio
New York University

Perry Lanier
Michigan State University

J. Michael Shaughnessy
Portland State University

Charles Vonder Embse
Central Michigan University

Special thanks to the students and teachers at these pilot schools!

Baker Demonstration School
Evanston, Illinois

Bertha Vos Elementary School
Traverse City, Michigan

Blair Elementary School
Traverse City, Michigan

Bloomfield Hills Middle School
Bloomfield Hills, Michigan

Brownell Elementary School
Flint, Michigan

Catlin Gabel School
Portland, Oregon

Cherry Knoll Elementary School
Traverse City, Michigan

Cobb Middle School
Tallahassee, Florida

Courtade Elementary School
Traverse City, Michigan

DeVeaux Junior High School
Toledo, Ohio

Duke School for Children
Durham, North Carolina

East Junior High School
Traverse City, Michigan

Eastern Elementary School
Traverse City, Michigan

Eastlake Elementary School
Chula Vista, California

Eastwood Elementary School
Sturgis, Michigan

Elizabeth City/Pasquotank Middle School
Elizabeth City, North Carolina

Franklinton Elementary School
Franklinton, North Carolina

Frick International Studies Academy
Pittsburgh, Pennsylvania

Gundry Elementary School
Flint, Michigan

Hawkins Elementary School
Toledo, Ohio

Hilltop Middle School
Chula Vista, California

Holmes Middle School
Flint, Michigan

Interlochen Elementary School
Traverse City, Michigan

Los Altos Elementary School
San Diego, California

Louis Armstrong Middle School
East Elmhurst, New York

McTigue Junior High School
Toledo, Ohio

National City Middle School
National City, California

Norris Elementary School
Traverse City, Michigan

Northeast Middle School
Minneapolis, Minnesota

Oak Park Elementary School
Traverse City, Michigan

Old Mission Elementary School
Traverse City, Michigan

Old Orchard Elementary School
Toledo, Ohio

Portland Middle School
Portland, Michigan

Reizenstein Middle School
Pittsburgh, Pennsylvania

Sabin Elementary School
Traverse City, Michigan

Shepherd Middle School
Shepherd, Michigan

Sturgis Middle School
Sturgis, Michigan

Terrell Lane Middle School
Louisburg, North Carolina

Tierra del Sol Middle School
Lakeside, California

Traverse Heights Elementary School
Traverse City, Michigan

University Preparatory Academy
Seattle, Washington

Washington Middle School
Vista, California

Waverley East Intermediate School
Lansing, Michigan

Waverly Middle School
Lansing, Michigan

West Junior High School
Traverse City, Michigan

Willlow Hill Elementary School
Traverse City, Michigan

Contents

Mathematical Highlights ... 4
The Unit Project: My Special Number ... 5

Investigation 1: The Factor Game ... 6
   1.1 Playing the Factor Game ... 6
   1.2 Playing to Win the Factor Game ... 10
   Applications—Connections—Extensions ... 12
   Mathematical Reflections ... 16

Investigation 2: The Product Game ... 17
   2.1 Playing the Product Game ... 17
   2.2 Making Your Own Product Game ... 19
   2.3 Classifying Numbers ... 20
   Applications—Connections—Extensions ... 22
   Mathematical Reflections ... 25

Investigation 3: Factor Pairs ... 26
   3.1 Arranging Space ... 26
   3.2 Finding Patterns ... 27
   3.3 Reasoning with Odd and Even Numbers ... 28
   Applications—Connections—Extensions ... 30
   Mathematical Reflections ... 35

Investigation 4: Common Factors and Multiples ... 36
   4.1 Riding Ferris Wheels ... 36
   4.2 Looking at Locust Cycles ... 38
   4.3 Planning a Picnic ... 39
   Applications—Connections—Extensions ... 40
   Mathematical Reflections ... 45

Investigation 5: Factorizations ... 46
   5.1 Searching for Factor Strings ... 46
   5.2 Finding the Longest Factor String ... 48
   5.3 Using Prime Factorizations ... 50
   Applications—Connections—Extensions ... 52
   Mathematical Reflections ... 57

Investigation 6: The Locker Problem ... 58
   6.1 Unraveling the Locker Problem ... 58
   Applications—Connections—Extensions ... 61
   Mathematical Reflections ... 64

The Unit Project: My Special Number ... 65

Prime Time

Why is time measured using 60 seconds in a minute (not 50 or 100), 60 minutes in an hour, and 24 hours in a day (not 23 or 25)?

Insects called cicadas spend most of their lives underground. Many come above ground only every 13 years or 17 years. In North America, many people call these cicadas 13-year locusts and 17-year locusts. Why are there no 12-year, 14-year, or 16-year locusts?

Why does your birthday fall on a different day of the week from one year to the next? Why is the same pattern also true for New Year's Day and the Fourth of July?

Everyone uses numbers. Think about the ways you can use them—for counting, for measuring, for making decisions. Numbers help you communicate, find information, use technology, and make purchases. Numbers also can help you think about situations like those on the opposite page.

Whole numbers have interesting properties and structures you may not have thought about before. Some numbers can be divided by many numbers, while others can be divided by only a few. In *Prime Time*, you will learn how to use these ideas about the structure of numbers to explain some curious patterns and to solve some interesting problems including the three on the opposite page.

Mathematical Highlights

In *Prime Time*, you will explore important new concepts about whole numbers and take a deeper look at some concepts you may already have encountered.

- Playing and analyzing the Factor Game and the Product Game help you learn about factors and multiples. As you come up with winning strategies, properties of prime numbers and composite numbers play an important part.

- What you know about factors and multiples helps you develop strategies to create your own Product Game.

- Making tile rectangles allows you to represent numbers and their factors. Finding patterns in the rectangles helps you visualize prime, composite, and square numbers.

- Determining whether a number is odd or even seems simple, but is an important part of thinking about numbers. As you make and prove conjectures about what happens when you add or multiply even and odd numbers, you build more ideas about numbers.

- As you solve interesting problems that require you to find the factors and multiples two or more numbers have in common, you see how these ideas relate to the real world.

- Playing the Product Puzzle lets you explore the factor strings of a number. Finding longer and longer factor strings leads you to discover a very important mathematical theorem.

- As you try to find the greatest common factor and least common multiple of a number, you see that you need to use the longest factor strings.

The Unit Project

My Special Number

Many people have a number they find interesting. Choose a whole number between 10 and 100 that you especially like.

In your journal

- record your number
- explain why you chose that number
- list three or four mathematical things about your number
- list three or four connections you can make between your number and your world

As you work through the investigations in *Prime Time,* you will learn lots of things about numbers. Think about how these new ideas apply to your special number, and add any new information about your number to your journal. You may want to designate one or two "special number" pages in your journal, where you can record this information. At the end of the unit, your teacher will ask you to find an interesting way to report to the class about your special number.

INVESTIGATION 1

The Factor Game

Today Jamie is 12 years old. Jamie has three younger cousins: Cam, Emilio, and Ester. They are 2, 3, and 8 years old respectively. The following mathematical sentences show that Jamie is

6 times as old as Cam,   4 times as old as Emilio,   and $1\frac{1}{2}$ times as old as Ester

$12 = 6 \times 2$ $12 = 4 \times 3$ $12 = 1\frac{1}{2} \times 8$

Notice that each of the whole numbers 2, 3, 4, and 6 can be multiplied by another whole number to get 12. We call 2, 3, 4, and 6 *whole number factors* or *whole number divisors* of 12. Although 8 is a whole number, it is not a whole number factor of 12, since we cannot multiply it by another whole number to get 12. To save time, we will simply use the word **factor** to refer to whole number factors.

1.1 Playing the Factor Game

The Factor Game is a two-person game in which players find factors of numbers on a game board. To play the game you will need Labsheet 1.1 and colored pens, pencils, or markers.

> **Problem 1.1**
>
> Play the Factor Game several times with a partner. Take turns making the first move. Look for moves that give the best scores. In your journal, record any strategies you find that help you to win.

■ **Problem 1.1 Follow-Up**

Talk with your partner about the games you played. Be prepared to tell the class about a good idea you discoved for playing the game well.

The Factor Game

1	2	3	4	5
6	7	8	9	10
11	12	13	14	15
16	17	18	19	20
21	22	23	24	25
26	27	28	29	30

Factor Game Rules

1. Player A chooses a number on the game board and circles it.
2. Using a different color, Player B circles all the proper factors of Player A's number. The **proper factors** of a number are all the factors of that number, except the number itself. For example, the proper factors of 12 are 1, 2, 3, 4, and 6. Although 12 is a factor of itself, it is not a proper factor.
3. Player B circles a new number, and Player A circles all the factors of the number that are not already circled.
4. The players take turns choosing numbers and circling factors.
5. If a player circles a number that has no factors left that have not been circled, that player loses a turn and does not get the points for the number circled.
6. The game ends when there are no numbers remaining with uncircled factors.
7. Each player adds the numbers that are circled with his or her color. The player with the greater total is the winner.

A sample game is shown on the following pages.

Sample Game

The first column describes the moves the players make. The other columns show the game board and score after each move.

		Cathy	Keiko
Cathy circles 24. Keiko circles 1, 2, 3, 4, 6, 8, and 12—the proper factors of 24.	The Factor Game ① ② ③ ④ 5 ⑥ 7 ⑧ 9 10 11 ⑫ 13 14 15 16 17 18 19 20 21 22 23 ㉔ 25 26 27 28 29 30	24	36
Keiko circles 28. Cathy circles 7 and 14—the factors of 28 that are not already circled.	The Factor Game ① ② ③ ④ 5 ⑥ ⑦ ⑧ 9 10 11 ⑫ 13 ⑭ 15 16 17 18 19 20 21 22 23 ㉔ 25 26 27 ㉘ 29 30	24 21	36 28
Cathy circles 27. Keiko circles 9—the only factor of 27 that is not already circled.	The Factor Game ① ② ③ ④ 5 ⑥ ⑦ ⑧ ⑨ 10 11 ⑫ 13 ⑭ 15 16 17 18 19 20 21 22 23 ㉔ 25 26 ㉗ ㉘ 29 30	24 21 27	36 28 9
Keiko circles 30. Cathy circles 5, 10, and 15—the factors of 30 that are not already circled.	The Factor Game ① ② ③ ④ ⑤ ⑥ ⑦ ⑧ ⑨ ⑩ 11 ⑫ 13 ⑭ ⑮ 16 17 18 19 20 21 22 23 ㉔ 25 26 ㉗ ㉘ 29 ㉚	24 21 27 30	36 28 9 30
Cathy circles 25. All the factors of 25 are circled. Cathy loses a turn and does not receive any points for this turn.	The Factor Game ① ② ③ ④ ⑤ ⑥ ⑦ ⑧ ⑨ ⑩ 11 ⑫ 13 ⑭ ⑮ 16 17 18 19 20 21 22 23 ㉔ ㉕ 26 ㉗ ㉘ 29 ㉚	24 21 27 30	36 28 9 30

		Cathy	Keiko
Keiko circles 26. Cathy circles 13—the only factor of 26 that is not circled.	The Factor Game board with 1, 6, 12, 24, 25, 26, 30 circled	24 21 27 30 13	36 28 9 30 26
		Cathy	Keiko
Keiko circles 22. Cathy circles 11—the only factor of 22 that is not circled.	The Factor Game board with 1, 6, 11, 12, 22, 24, 25, 26, 30 circled	24 21 27 30 13 11	36 28 9 30 26 22
		Cathy	Keiko
No numbers remain with uncircled factors. Keiko wins the game.	The Factor Game board with 1, 6, 11, 12, 22, 24, 25, 26, 30 circled	24 21 27 30 13 11	36 28 9 30 26 22
	Total	126	151

1.2 Playing to Win the Factor Game

Did you find that some numbers are better than others to pick for the first move in the Factor Game? For example, if you pick 22, you get 22 points and your opponent gets only $1 + 2 + 11 = 14$ points. However, if you pick 18, you get 18 points, and your opponent gets $1 + 2 + 3 + 6 + 9 = 21$ points!

Make a table of all the possible first moves (numbers from 1 to 30) you could make. For each move, list the proper factors, and record the scores you and your opponent would receive. Your table might start like this:

First move	Proper factors	My score	Opponent's score
1	none	lose a turn	0
2	1	2	1
3	1	3	1
4	1, 2	4	3

Problem 1.2

Use your list to figure out the best and worst first moves.

A. What is the best first move? Why?

B. What is the worst first move? Why?

C. Look for other patterns in your list. Describe an interesting pattern that you find.

■ **Problem 1.2 Follow-Up**

1. List all the first moves that allow your opponent to score only one point. These kinds of numbers have a special name. They are called **prime numbers**.

2. Are all prime numbers good first moves? (A number is a good first move if the player picking the number scores more points than his or her opponent.) Explain your answer.

3. List all the first moves that allow your opponent to score more than one point. These kinds of numbers also have a special name. They are called **composite numbers**.

4. Are composite numbers good first moves? Explain your answer.

5. Which first move would make you lose a turn? Why?

Did you know?

The search for prime numbers has fascinated mathematicians for a very long time. We know that there are an infinite number of primes, but we have no way to predict which numbers are prime. We must test each number to see if it has exactly two factors—1 and itself. For very large numbers, this testing takes a long time, even with the help of a supercomputer that can perform 16 billion calculations per second!

In 1994, David Slowinski, a computer scientist at Cray Research, found a prime number with 258,716 digits. The previous record holder had 227,832 digits. Large prime numbers are of special importance in coding systems for transmitting secret information. The difficulty of breaking these codes depends on the difficulty of factoring a composite number with 100 or more digits into prime factors with at least 50 digits. Computer programmers think that such a problem would require over a billion years on the largest imaginable supercomputer.

Adapted from Phillips et al., *Addenda Series, Grades 5–8: Patterns and Functions* (Reston, Va.: National Council of Teachers of Mathematics, 1991), p. 21, and information provided by Cray Research, Inc.

Applications • Connections • Extensions

As you work on these ACE questions, use your calculator whenever you need it.

Applications

1. Your opponent in the Factor Game claims that 6 is a factor of 24. How can you check to see whether this is correct?

2. What factor is paired with 6 to give 24?

3. What factor is paired with 5 to give 45?

4. What factor is paired with 3 to give 24?

5. What factor is paired with 6 to give 45?

6. What factor is paired with 6 to give 48?

7. What factor is paired with 11 to give 121?

8. What factor is paired with 12 to give 48?

9. What factor is paired with 11 to give 111?

10. The Factor Game can be played on a 49-board, which contains whole numbers from 1 to 49.

The Factor Game

1	2	3	4	5	6	7
8	9	10	11	12	13	14
15	16	17	18	19	20	21
22	23	24	25	26	27	28
29	30	31	32	33	34	35
36	37	38	39	40	41	42
43	44	45	46	47	48	49

a. Extend your table for analyzing first moves on a 30-factor game board to include all the numbers on a 49-board.

b. What new primes do you find?

11. Suppose your opponent has the first move on the 49-board and chooses 49.
   a. How many points does your opponent score for this round?
   b. How many points do you score for this round?

12. What is the best first move on a 49-board? Why?

13. What is the worst first move on a 49-board? Why?

14. **a.** What do you get when you use your calculator to divide 84 by 14? What does this tell you about 14 and 84?
   b. What do you get when you use your calculator to divide 84 by 15? What does this tell you about 15 and 84?

15. Use the ideas from this investigation to list at least five facts about the number 30.

16. What is my number?
   Clue 1 My number has two digits, and both digits are even.
   Clue 2 The sum of my number's digits is 10.
   Clue 3 My number has 4 as a factor.
   Clue 4 The difference between the two digits of my number is 6.

Connections

17. A class of 30 students is to be divided into equal-size groups. What group sizes are possible?

18. Long ago, people observed the sun rising and setting over and over at about equal intervals. They decided to use the amount of time between two sunrises as the length of a day. They divided the day into 24 hours. Use what you know about factors to answer these questions:

   a. Why is 24 a more convenient choice than 23 or 25?
   b. If you were to select a number different from 24 to represent the hours in a day, what number would you choose? Why?

Extensions

19. Suppose you and a friend decide to use a 100-board to play the Factor Game.
   a. What would your score be if your friend chose 100 as the first move?
   b. What would your score be if your friend chose 99 as the first move?
   c. What is the best first move?

20. The sum of the proper factors of a number may be greater than, less than, or equal to the number. Ancient mathematicians used this idea to classify numbers as **abundant, deficient,** and **perfect.** Each whole number greater than 1 falls into one of these three categories.
   a. Draw and label three circles as shown below. The numbers 12, 15, and 6 have been placed in the appropriate circles. Use your factor list to figure out what each label means. Then, write each whole number from 2 to 30 in the correct circle.

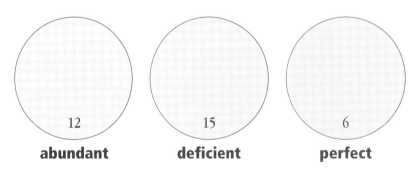

   abundant **deficient** **perfect**

   b. Do the labels seem appropriate? Why or why not?
   c. In which circle would 36 belong?
   d. In which circle would 55 belong?

21. **a.** If you choose 16 as a first move in the Factor Game, how many points does your opponent get? How does your opponent's score for this turn compare to yours?
   b. If you choose 4 as a first move, how many points does your opponent get? How does your opponent's score for this turn compare to yours?
   c. Find some other numbers that have the same pattern of scoring as 4 and 16. These numbers could be called **near-perfect numbers.** Why do you think this name fits?

Did you know?

Is there a largest perfect number? Mathematicians have been trying for hundreds of years to find the answer to this question. You might like to know that the next largest perfect number, after 6 and 28, is 496!

Mathematical Reflections

In Investigation 1, you played and analyzed the Factor Game. These questions will help you summarize what you have learned:

1. Which numbers are good first moves? What makes these numbers good moves?
2. Which numbers are bad first moves? What makes these numbers bad moves?
3. What did your analysis of the factor game tell you about prime numbers?

Think about your answers to these questions, discuss your ideas with other students and your teacher, and then write a summary of your findings in your journal.

Have you remembered to write about your special number?

INVESTIGATION 2

The Product Game

In the Factor Game, you start with a number and find its factors. In the Product Game, you start with factors and find their product. The diagram shows the relationship between factors and their product.

2.1 Playing the Product Game

The Product Game board consists of a list of factors and a grid of products. Two players compete to get four squares in a row—up and down, across, or diagonally. To play the Product Game, you will need Labsheet 2.1, two paper clips, and colored markers or game chips. The rules for the Product Game are given on the next page.

The Product Game

1	2	3	4	5	6
7	8	9	10	12	14
15	16	18	20	21	24
25	27	28	30	32	35
36	40	42	45	48	49
54	56	63	64	72	81

Factors:
1 2 3 4 5 6 7 8 9

Investigation 2: The Product Game 17

Problem 2.1

Play the Product Game several times with a partner. Look for interesting patterns and winning strategies. Make notes of your observations.

Product Game Rules

1. Player A puts a paper clip on a number in the factor list. Player A does not mark a square on the product grid because only one factor has been marked; it takes two factors to make a product.
2. Player B puts the other paper clip on any number in the factor list (including the same number marked by Player A) and then shades or covers the product of the two factors on the product grid.
3. Player A moves *either one* of the paper clips to another number and then shades or covers the new product.
4. Each player, in turn, moves a paper clip and marks a product. If a product is already marked, the player does not get a mark for that turn. The winner is the first player to mark four squares in a row—up and down, across, or diagonally.

Problem 2.1 Follow-Up

1. Suppose one of the paper clips is on 5. What products can you make by moving the other paper clip?

The products you listed in question 1 are multiples of 5. A **multiple** of a number is the product of that number and another whole number.

If a number is a multiple of 5, then 5 is a factor of that number. These four sentences are all ways of expressing $5 \times 3 = 15$:

5 is a factor of 15.
3 is a factor of 15.
15 is a multiple of 5.
15 is a multiple of 3.

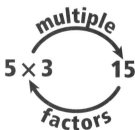

2. List five multiples of 5 that are not on the game board.
3. Suppose one of the paper clips is on 3. What products can you make by moving the other paper clip?
4. List five multiples of 3 that are not on the game board.

2.2 Making Your Own Product Game

Suppose you want to create a product game that takes less time to play or, perhaps, more time to play than the game with the 6 × 6 product grid. You would have to decide what numbers to include in the factor list and what products to include in the product grid.

Problem 2.2

Work with your partner to design a new game board for the Product Game.
- Choose factors to include in your factor list.
- Determine the products you need to include on the game board.
- Find a game board that will accommodate all the products.
- Decide how many squares a player must get in a row—up and down, across, or diagonally—to win.

Make the game board. Play your game against your partner; then make any changes you both agree would make your game better.

Switch game boards with another pair, and play their game. Give them some written suggestions about how they can improve their game. Read the suggestions for improving your game, then make any changes you and your partner think are necessary.

■ Problem 2.2 Follow-Up

Write a paragraph about why you think your game board is interesting to use for playing the Product Game. In the paragraph, describe any problems you ran into while making the board, and explain how you solved them.

2.3 Classifying Numbers

Now that you know how to find the factors and multiples of a number, you can explore how the factors and multiples of two or more numbers are related. Venn diagrams are useful tools for exploring these relationships. A **Venn diagram** uses circles to show things that belong together. For example, the Venn diagram below shows one way to group whole numbers.

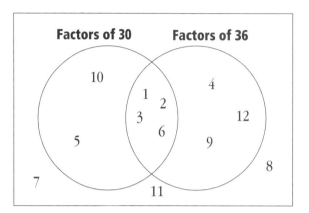

One circle represents all the whole numbers that are factors of 30. The other circle represents all the whole numbers that are factors of 36. The first 12 whole numbers have been placed in the correct regions of the diagram. Notice that the numbers that are not factors of 30 or 36 lie outside the circles. Why do the circles intersect (overlap)? What do the numbers in the intersection have in common?

Problem 2.3

Copy the Venn diagram below.

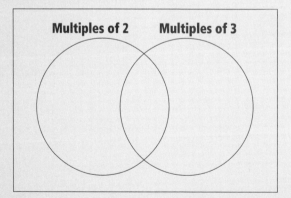

Find at least five numbers that belong in each region of the diagram. Think about what numbers belong in the intersection of the circles and what numbers belong outside of the circles.

■ Problem 2.3 Follow-Up

1. What factors do the numbers in the intersection of the circles have in common?
2. Add a new circle to the diagram with the label "Multiples of 5," as shown below. Replace your numbers in the correct regions, and make sure you have at least two numbers in each part of the diagram.

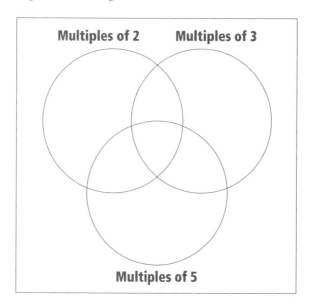

Investigation 2: The Product Game

Applications • Connections • Extensions

As you work on these ACE questions, use your calculator whenever you need it.

Applications

1. Marena just marked 18 on the 6 × 6 Product Game board. On which factors might the paper clips be? List all the possibilities.

2. Find two products on the board, other than 18, that can be made in more than one way. List all the factor pairs that give each product.

3. On the 6 × 6 Product Game board, 81 is a multiple of which factors?

4. On the 6 × 6 Product Game board, suppose your markers are on 16, 18, and 28, and your opponent's markers are on 14, 21, and 30. The paper clips are on 5 and 6. It is your turn to move a paper clip.

The Product Game

1	2	3	4	5	6
7	8	9	10	12	**14**
15	**16**	**18**	20	**21**	24
25	27	**28**	**30**	32	35
36	40	42	45	48	49
54	56	63	64	72	81

Factors:

1 2 3 4 5 6 7 8 9

  a. List the possible moves you could make.
  b. Which move(s) would give you three markers in a row?
  c. Which move(s) would allow you to block your opponent?
  d. Which move would you make? Explain your strategy.

In 5–8, find two numbers that can be multiplied to give each product. Do not use 1 as one of the numbers.

5. 84 **6.** 145 **7.** 250 **8.** 300

22 Prime Time

9. What factors were used to create this Product Game board?

4	6	14
9	21	49

Factors: ___ ___ ___

10. What factors were used to create this Product Game board? What number is missing from the grid?

9	15	18	■
21	?	30	35
■	36	42	49

Factors: ___ ___ ___ ___

11. Draw and label a Venn diagram in which one circle represents multiples of 3 and another circle represents multiples of 5. Place the multiples of 3 and the multiples of 5 from 1 to 60 in the appropriate regions of the diagram. The numbers that are multiples of both 3 and 5 are the **common multiples** of 3 and 5. These numbers go in the intersection of the two circles.

12. Find at least five numbers that belong in each of the regions of this Venn diagram.

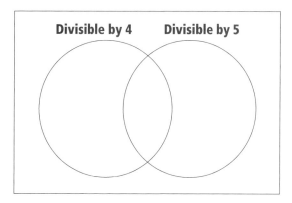

Connections

13. What numbers is 36 a multiple of?

14. Using the words *factor, divisor, multiple, product,* and *divisible by,* write as many statements as you can about the mathematical sentence $7 \times 9 = 63$.

15. Draw and label a Venn diagram in which one circle contains the **divisors** (factors) of 42 and the other contains the divisors of 60. The divisors of both 42 and 60 are the **common factors** of the two numbers. The common factors should go in the intersection of the two circles.

16. Find all the common multiples of 4 and 11 that are less than 100.

17. The cast of the school play had a party at the drama teacher's house. There were 20 cookies and 40 carrot sticks served as refreshments. Each cast member had the same number of whole cookies and the same number of whole carrot sticks, and nothing was left over. The drama teacher did not eat. How many cast members might have been at the party? Explain your answer.

18. A restaurant is open 24 hours a day. The manager wants to divide the day into workshifts of equal length. Show the different ways this can be done. The shifts should not overlap, and all shifts should be a whole number of hours long.

19 a. In developing ways to calculate time, astronomers divided an hour into 60 minutes. Why is 60 a good choice (better than 59 or 61)?

b. If you were to select another number to represent the minutes in an hour, what would be a good choice? Why?

Extensions

20. What is my number?
Clue 1 When you divide my number by 5, the remainder is 4.
Clue 2 My number has two digits, and both digits are even.
Clue 3 The sum of the digits is 10.

Mathematical Reflections

In this investigation, you played and analyzed the Product Game. These questions will help you summarize what you have learned:

1) In the Product Game, describe the relationship between the numbers in the factor list and the products in the grid.

2) What are the multiples of a number and how do you find them?

3) Using the words *factor, divisor, multiple,* and *divisible by,* write as many statements as you can about this mathematical sentence:

$$4 \times 7 = 28$$

Think about your answers to these questions, discuss your ideas with other students and your teacher, and then write a summary of your findings in your journal.

Write something new that you have learned about your special number now that you have played the Factor Game and the Product Game. Would your special number be a good first move in either game? Why or why not?

INVESTIGATION 3

Factor Pairs

In the Factor Game and the Product Game, you found that factors come in pairs. Once you know one factor of a number, you can find another factor. For example, 3 is a factor of 12, and because $3 \times 4 = 12$, you know 4 is also a factor of 12. In this investigation, you will look at factor pairs in a different way.

3.1 Arranging Space

Every year, Meridian Park has an exhibit of arts and crafts. People who want to exhibit their work rent a space for $20 per square yard. All exhibit spaces must have a rectangular shape. The length and width of an exhibit space must be whole numbers of yards.

Problem 3.1

Terrapin Crafts wants to rent a space of 12 square yards.

A. Use 12 square tiles to represent the 12 square yards. Find all the possible ways the Terrapin Crafts owner can arrange the squares. Copy each rectangle you make onto grid paper, and label it with its dimensions (length and width).

B. How are the rectangles you found and the factors of 12 related?

Suppose Terrapin Crafts decided it wanted a space of 16 square yards.

C. Find all the possible ways the Terrapin Crafts owner can arrange the 16 square yards. Copy each rectangle you make onto grid paper, and label it with its dimensions.

D. How are these rectangles and the factors of 16 related?

26 Prime Time

■ **Problem 3.1 Follow-Up**

What factors do 12 and 16 have in common?

3.2 Finding Patterns

Will likes to find number patterns. He wonders if there are any interesting patterns in the rectangles that can be made for the numbers from 1 to 30.

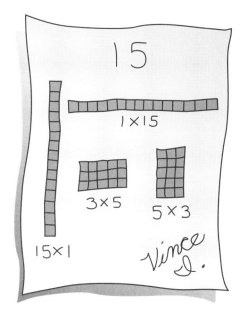

In this problem, your class will make rectangles for all the whole numbers from 1 through 30. When all the rectangles are displayed, you can look for interesting patterns.

Work with a partner or a small group so that you can check each other's work. With your teacher, decide which numbers your group will be responsible for.

Problem 3.2

Work with your group to decide how to divide up the work for the numbers you have been assigned.

Cut out a grid-paper model of each rectangle you can make for each of the numbers you have been assigned. You may want to use tiles to help you find the rectangles.

Write each number at the top of a sheet of paper, and tape all the rectangles for that number to the sheet. Display the sheets of rectangles in order from 1 to 30 around the room.

When all the numbers are displayed, look for patterns. Be prepared to discuss patterns you find with your classmates.

■ **Problem 3.2 Follow-Up**
1. Which numbers have the most rectangles? What kind of numbers are these?
2. Which numbers have the fewest rectangles? What kind of numbers are these?
3. Which numbers are **square numbers** (numbers whose tiles can be arranged to form a square)?
4. If you know the rectangles you can make for a number, how can you use this information to list the factors of the number? Use an example to show your thinking.

3.3 Reasoning with Odd and Even Numbers

An **even number** is a number that has 2 as a factor. An **odd number** is a number that does not have 2 as a factor. In this problem, you will study patterns involving odd and even numbers. First, you will learn a way of modeling odd and even numbers. Then, you will make conjectures about sums and products of odd and even numbers. A *conjecture* is your best guess about a relationship. You can use the models to justify, or prove, your conjectures.

"AN ODD NUMBER"

Will's friend, Jocelyn, makes models for whole numbers by arranging square tiles in a special pattern. Here are Jocelyn's tile models for the numbers from 1 to 7.

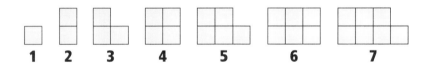

Discuss with your class how the models of even numbers are different from the models of odd numbers. Then describe the models for 50 and 99.

Problem 3.3

Make a conjecture about whether each result below will be even or odd. Then use tile models or some other method to justify your conjecture.

A. The sum of two even numbers

B. The sum of two odd numbers

C. The sum of an odd number and an even number

D. The product of two even numbers

E. The product of two odd numbers

F. The product of an odd number and an even number

Problem 3.3 Follow-Up

1. Is 0 an even number or an odd number? Explain your answer.
2. Without building a tile model, how can you tell whether a sum of numbers—such as 127 + 38—is even or odd?

Applications • Connections • Extensions

As you work on these ACE questions, use your calculator whenever you need it.

Applications

In 1–6, give the dimensions of each rectangle that can be made from the given number of tiles. Then, use the dimensions of the rectangles to list all the factor pairs for each number.

1. 24   **2.** 32   **3.** 48

4. 45   **5.** 60   **6.** 72

In 7 and 8, write a description, with examples, of numbers that have the given factors.

7. exactly two factors   **8.** an odd number of factors

9. Lupe has chosen a mystery number. His number is larger than 12 and smaller than 40, and it has exactly three factors. What could his number be? Use the displays of rectangles for the numbers 1 to 30 to help you find Lupe's mystery number. You may need to think about what the displays for the numbers 31 to 40 would look like.

10. Without building a tile model, how can you tell whether a sum of numbers—such as 13 + 45 + 24 + 17 is even or odd?

In 11–14, make a conjecture about whether each result will be odd or even. Use tiles, a picture, or some other way to justify your conjectures.

11. An even number minus and even number

12. An odd number minus an odd number

13. An even number minus an odd number

14. An odd number minus an even number

15. How can you tell whether a number is even or odd? Explain or illustrate your answer in at least two ways.

Connections

16. a. List all the numbers less than or equal to 50 that are divisible by 5.

b. Describe a pattern you see in your list that you can use to determine whether a large number—such as 1,276,549—is divisible by 5.

c. Which numbers in your list are divisible by 2?

d. Which numbers in your list are divisible by 10?

e. How do the lists in parts c and d compare? Why does this result make sense?

17. A group of students designs card displays for football games. They use 100 square cards for each display. Each card contains part of a picture or message. At the game, 100 volunteers hold up the cards to form a complete picture. The students have found that the pictures are most effective if the volunteers sit in a rectangular arrangement. What seating arrangements are possible? Which would you choose? Why?

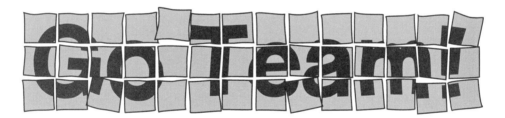

18. The school band has 64 members. The band marches in the form of a rectangle. What rectangles can the band director make by arranging the members of the band? Which of these arrangements is most appealing to you? Why?

19. How many rectangles can you build with a prime number of square tiles?

Extensions

20. Find three numbers you can multiply to get 300.

In 21–23, tell whether each number is a square number. Justify your answer.

21. 196   **22.** 225   **23.** 360

24. a. Find at least five numbers that belong in each region of the Venn diagram below.

   b. What do the numbers in the intersection have in common?

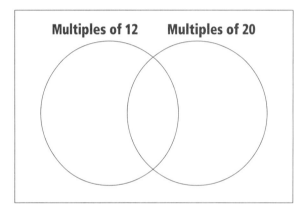

25. a. Below is the complete list of the proper factors of a certain number. What is the number?

   1, 2, 3, 4, 6, 7, 12, 14, 21, 28, 42, 49, 84, 98, 147, 196, 294

   b. List each of the factor pairs for the number.

   c. What rectangles could be made to show the number?

26. For any three consecutive numbers (whole numbers in a row), such as 31, 32, 33, or 52, 53, 54, what is true about odds and evens? Explain your thinking.

27. Ji Young conjectured that, in every three consecutive whole numbers, one number will be divisible by 3. Do you think Ji Young is correct? Explain.

28. How many consecutive numbers do you need to guarantee that one of the numbers is divisible by 5?

29. How many consecutive numbers do you need to guarantee that one of the numbers is divisible by 6?

30. Choose a nonprime number between 900 and 1000, and find all of the factors of the number. The chart on the next page will help you select an interesting number.

Factor Counts ∎   Each * Stands for a Factor

900	`***************************`	951	`****`
901	`****`	952	`****************`
902	`********`	953	`**`
903	`********`	954	`************`
904	`********`	955	`****`
905	`****`	956	`******`
906	`********`	957	`********`
907	`**`	958	`****`
908	`******`	959	`****`
909	`******`	960	`********************************`
910	`****************`	961	`***`
911	`**`	962	`********`
912	`********************`	963	`******`
913	`****`	964	`******`
914	`****`	965	`****`
915	`********`	966	`****************`
916	`******`	967	`**`
917	`****`	968	`************`
918	`****************`	969	`********`
919	`**`	970	`********`
920	`****************`	971	`**`
921	`****`	972	`******************`
922	`****`	973	`****`
923	`****`	974	`****`
924	`************************`	975	`************`
925	`******`	976	`**********`
926	`****`	977	`**`
927	`******`	978	`********`
928	`************`	979	`****`
929	`**`	980	`******************`
930	`****************`	981	`******`
931	`******`	982	`****`
932	`******`	983	`**`
933	`****`	984	`****************`
934	`****`	985	`****`
935	`********`	986	`********`
936	`************************`	987	`********`
937	`**`	988	`************`
938	`********`	989	`****`
939	`****`	990	`************************`
940	`************`	991	`**`
941	`**`	992	`************`
942	`********`	993	`****`
943	`****`	994	`********`
944	`**********`	995	`****`
945	`****************`	996	`************`
946	`********`	997	`**`
947	`**`	998	`****`
948	`************`	999	`********`
949	`****`	1000	`****************`
950	`************`		

Prime Time

Mathematical Reflections

In this investigation, you analyzed factor pairs. You found that factor pairs for a number are related to the rectangles that can be made from that number of square tiles. You also investigated even and odd numbers. These questions will help you summarize what you have learned:

1 Explain how the rectangles you can make using 24 tiles are related to the factor pairs of the number 24.

2 Summarize what you know about the sums and products of odd and even numbers. Justify your statements.

3 How can you tell if a number is divisible by 2? By 5? By 10?

Think about your answers to these questions, discuss your ideas with other students and your teacher, and then write a summary of your findings in your journal.

Write about your special number! What can you say about your number now? Is it even? Is it odd? How many factor pairs does it have?

INVESTIGATION 4

Common Factors and Multiples

There are many things in the world that happen over and over again in set cycles. Sometimes we want to know when two things with different cycles will happen at the same time. Knowing about factors and multiples can help you to solve such problems.

Let's start by comparing the multiples of 20 and 30.

- The multiples of 20 are 20, 40, 60, 80, 100, 120, . . .
- The multiples of 30 are 30, 60, 90, 120, 150, 180, . . .

The numbers 60, 120, 180, 240, . . . are multiples of both 20 and 30. We call these numbers **common multiples** of 20 and 30.

Now let's compare the factors of 12 and 30.

- The factors of 12 are 1, 2, 3, 4, 6, and 12.
- The factors of 30 are 1, 2, 3, 5, 6, 10, 15, and 30.

The numbers 1, 2, 3, and 6 are factors of both 12 and 30. We call these numbers **common factors** of 12 and 30.

4.1 Riding Ferris Wheels

One of the most popular rides at a carnival or amusement park is the Ferris wheel.

> **Did you know?**
>
> The largest Ferris Wheel was built for the World's Columbian Exposition in Chicago in 1893. The wheel could carry 2160 people in its 36 passenger cars. Can you figure out how many people could ride in each car?

Problem 4.1

You and your little sister go to a carnival that has both a large and a small Ferris wheel. You get on the large Ferris wheel at the same time your sister gets on the small Ferris wheel. The rides begin as soon as you are both buckled into your seats. Determine the number of seconds that will pass before you and your sister are both at the bottom again

A. if the large wheel makes one revolution in 60 seconds and the small wheel makes one revolution in 20 seconds.

B. if the large wheel makes one revolution in 50 seconds and the small wheel makes one revolution in 30 seconds.

C. if the large wheel makes one revolution in 10 seconds and the small wheel makes one revolution in 7 seconds.

Problem 4.1 Follow-Up

For parts A–C in Problem 4.1, determine the number of times each Ferris wheel goes around before you and your sister are both on the ground again.

4.2 Looking at Locust Cycles

Cicadas spend most of their lives underground. Some cicadas—commonly called 13-year locusts—come above ground every 13 years, while others—called 17-year locusts—come out every 17 years.

> **Problem 4.2**
>
> Stephan's grandfather told him about a terrible year when the cicadas were so numerous that they ate all the crops on his farm. Stephan conjectured that both 13-year and 17-year locusts came out that year. Assume Stephan's conjecture is correct.
>
> **A.** How many years pass between the years when both 13-year and 17-year locusts are out at the same time? Explain how you got your answer.
>
> **B.** Suppose there were 12-year, 14-year, and 16-year locusts, and they all came out this year. How many years will it be before they all come out together again? Explain how you got your answer.

■ **Problem 4.2 Follow-Up**

For parts A and B of Problem 4.2, tell whether the answer is less than, greater than, or equal to the product of the locust cycles.

"BELIEVE ME, THEY'RE NOT EXPECTING US. WE'RE 387-YEAR LOCUSTS."

©1991 by Sidney Harris. From *You Want Proof? I'll Give You Proof!*
W.H. Freeman, New York.

4.3 Planning a Picnic

Common factors and common multiples can be used to figure out how many people can share things equally.

Problem 4.3

Miriam's uncle donated 120 cans of juice and 90 packs of cheese crackers for the school picnic. Each student is to receive the same number of cans of juice and the same number of packs of crackers.

What is the largest number of students that can come to the picnic and share the food equally? How many cans of juice and how many packs of crackers will each student receive? Explain how you got your answers.

Problem 4.3 Follow-Up

If Miriam's uncle eats two packs of crackers before he sends the supplies to the school, what is the largest number of students that can come to the picnic and share the food equally? How many cans of juice and how many packs of crackers will each receive?

Investigation 4: Common Factors and Multiples

Applications • Connections • Extensions

As you work on these ACE questions, use your calculator whenever you need it.

Applications

In 1–4, list the common multiples between 1 and 100 for each pair of numbers. Then find the least common multiple for each pair.

1. 8 and 12

2. 3 and 15

3. 7 and 11

4. 9 and 10

In 5–7, find two pairs of numbers with the given number as their least common multiple.

5. 10

6. 36

7. 60

In 8–10, list the common factors for each pair of numbers. Then find the greatest common factor for each pair.

8. 18 and 30

9. 9 and 25

10. 60 and 45

In 11–13, find two pairs of numbers with the given number as their greatest common factor.

11. 8

12. 1

13. 15

Connections

14. Mr. Vicario and his 23 students are planning to have hot dogs at their class picnic. Hot dogs come in packages of 12, and hot dog buns come in packages of 8.

   a. What is the smallest number of packages of hot dogs and the smallest number of packages of buns Mr. Vicario can buy so that everyone including him gets the same number of hot dogs and buns and there are no leftovers? How many hot dogs and buns does each person get?

   b. If the class invites the principal, the secretary, the bus driver, and three parents to help supervise, how many packages of hot dogs and buns will Mr. Vicario need to buy? How many hot dogs and buns will each person get if there are to be no leftovers?

15. The school cafeteria serves pizza every sixth day and applesauce every eighth day. If pizza and applesauce are both on today's menu, how many days will it be before they are both on the menu again?

16. Two neon signs are turned on at the same time. Both signs blink as they are turned on. One sign blinks every 9 seconds. The other sign blinks every 15 seconds. In how many seconds will they blink together again?

Extensions

17. Stephan told his biology teacher his conjecture that the terrible year of the cicadas occurred because 13-year and 17-year locusts came out at the same time. The teacher thought Stephan's conjecture was probably incorrect, because cicadas in a particular area seem to be either all 13-year locusts or all 17-year locusts, but not both. Stephan read about cicadas and found out that they are eaten very quickly by lots of predators. However, the cicadas are only in danger if their cycle occurs at the same time as the cycles of their predators. Stephan suspects that the reason there are 13-year and 17-year locusts but not 12-year, 14-year, or 16-year locusts has to do with predator cycles.

   a. Suppose cicadas have predators with 2-year cycles. How often would 12-year locusts face their predators? Would life be better for 13-year locusts?

   b. Suppose 12-year and 13-year locusts have predators with both 2-year and 3-year cycles. Suppose both kinds of locusts and both kinds of predators came out this year. When would the 12-year locusts again have to face both kinds of predators at the same time? What about the 13-year locusts? Which type of locust do you think is better off?

18. Suppose that in some distant part of the universe there is a star with four orbiting planets. One planet makes a trip around the star in 6 earth years, the second planet takes 9 earth years, the third takes 15 earth years, and the fourth takes 18 earth years. Suppose that at some time the planets are lined up as pictured. This phenomenon is called *conjunction*. How many years will it take before the planets return to this position?

19. Examine the number pattern below. You can use the tiles to help you see a pattern.

   1   = 1
   1 + 3   = 4
   1 + 3 + 5   = 9
   1 + 3 + 5 + 7 = 16

   a. Complete the next four rows in the number pattern.

   b. What is the sum in row 20?

   c. In what row will the sum be 576? What is the last number in the sum in this row? Explain how you got your answers.

20. Examine the pattern below. Using tiles may help you see a pattern.

   2   = 2
   2 + 4   = 6
   2 + 4 + 6   = 12
   2 + 4 + 6 + 8 = 20

   a. Complete the next four rows in the pattern.

   b. What is the sum in row 20?

   c. In what row will the sum be 110? What is the last number in the sum in this row? Explain how you got your answers.

21. Ms. Soong has a lot of pens in her desk drawer. She says that if you divide the total number of pens by 2, 3, 4, 5, or 6, you get a remainder of 1. What is the smallest number of pens that could be in Ms. Soong's drawer?

22. What is the mystery number pair?
   Clue 1 The greatest common factor of the mystery pair is 7.
   Clue 2 The least common multiple of the mystery pair is 70.
   Clue 3 Both of the numbers in the mystery pair have two digits.
   Clue 4 One of the numbers in the mystery pair is odd and the other is even.

23. While Min Ji was reading through her old journals, she noticed that on March 31, 1993, she had written the date 3-31-93. It looked like a multiplication problem, $3 \times 31 = 93$. Find as many other such dates as you can.

Mathematical Reflections

In this investigation, you used the ideas of common factors and common multiples to help you solve problems. These questions will help you summarize what you have learned:

1) Look at the three problems in this investigation. For which problems was it helpful to find common multiples? For which problems was it helpful to find common factors?

2) Make up a word problem you can solve by finding common factors and a different problem you can solve by finding common multiples. Solve your problems, and explain how you know your answers are correct.

3) Describe how you can find the common factors for two numbers.

4) Describe how you can find the common multiples for two numbers.

Think about your answers to these questions, discuss your ideas with other students and your teacher, and then write a summary of your findings in your journal.

Don't forget to write about your special number!

INVESTIGATION 5

Factorizations

Some numbers can be written as the product of several different pairs of factors. For example, 100 can be written as 1×100, 2×50, 4×25, 5×20, and 10×10. It is also possible to write 100 as the product of three factors, such as $2 \times 2 \times 25$ and $2 \times 5 \times 10$. Can you find a longer string of factors with a product of 100?

5.1 Searching for Factor Strings

The Product Puzzle on Labsheet 5.1 is a number-search puzzle. Your task is to find strings of numbers with a product of 840.

The Product Puzzle

30	×	14	×	8	×	7	×	210	×
×	2	×	4	×	3	×	2	×	2
105	×	2	×	5	×	84	×	56	×
×	21	×	2	×	7	×	8	×	3
40	×	20	×	4	×	7	×	5	×
×	4	×	28	×	5	×	3	×	2
6	×	8	×	21	×	2	×	105	×
×	2	×	10	×	2	×	5	×	2
32	×	3	×	14	×	60	×	56	×
×	5	×	8	×	15	×	7	×	3

Strings Found in the Product Puzzle

105 × 2 × 4

Problem 5.1

In the Product Puzzle, find as many factor strings for 840 as you can. A string can go around corners as long as there is a multiplication sign, ×, between any two numbers. When you find a string, draw a loop around it. Keep a record of the strings you find.

Problem 5.1 Follow-Up
1. Name two strings with a product of 840 that are not in the puzzle.
2. What is the longest string you found?
3. If possible, name a string with a product of 840 that is longer than any string you found in the puzzle. Do not consider strings that contain 1.
4. How do you know when you have found the longest possible string of factors for a number?
5. How many distinct longest strings of factors are there for a given number? Strings are distinct if they are different in some way other than the order in which the factors are listed. Do not consider strings that contain 1.

5.2 Finding the Longest Factor String

The strings of factors for a number are called *factorizations* of the number. When you search for factorizations of large numbers, it helps to keep an orderly record of your steps. One way to do this is to make a *factor tree*.

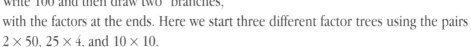

To find the longest factorization for 100, for example, you might proceed as follows:

Find two factors with a product of 100. Write 100 and then draw two "branches," with the factors at the ends. Here we start three different factor trees using the pairs 2×50, 25×4, and 10×10.

Where possible, break down each of the factors into the product of two factors. Write these factors in a new row of your tree. Draw branches to show how these factors are related to the numbers in the row above. The 2 in the first tree below does not break down any further, so we draw a single branch and repeat the 2 in the next row.

The numbers in the bottom row of the last two trees do not break down any further. These trees are complete. The bottom row of the first tree contains 25, so we complete this tree by breaking 25 into 5×5.

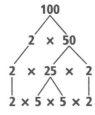

Notice that the bottom row of each tree contains the same factors, although the order of the factors is different. All three trees indicate that the longest factorization for 100 is $2 \times 2 \times 5 \times 5$. Think about why you cannot break this string down any further.

You can use a shortcut to write $2 \times 2 \times 5 \times 5$. In this shortcut notation, the string is written $2^2 \times 5^2$, which is read "2 squared times 5 squared." The small raised numbers are called exponents. An *exponent* tells us how many times the factor is repeated. For example, $2^2 \times 5^4$ means that the 2 is repeated twice and the 5 is repeated four times. So $2^2 \times 5^4$ is the same as $2 \times 2 \times 5 \times 5 \times 5 \times 5$.

Problem 5.2

Work with a partner to find the longest factorization for 600. You may make a factor tree or use another method. When you are finished, compare your results with the results of your classmates.

Did everyone produce the same results? If so, what was is the longest factorization for 600? If not, what differences occurred?

■ Problem 5.2 Follow-Up

1. Find the longest factorizations for 72 and 120.
2. What kinds of numbers are in the longest factor strings for the numbers you found?
3. How do you know that the factor strings you found cannot be broken down any further?
4. Rewrite the factor strings you found for 72, 120, and 600 using shortcut notation.

5.3 Using Prime Factorizations

In Investigation 4, you found common multiples and common factors of numbers by comparing lists of their multiples and factors. In this problem, you will explore a method for finding the *greatest common factor* and the *least common multiple* of two numbers by using their prime factorizations.

Heidi says she can find the greatest common factor and the least common multiple of a pair of numbers by using their prime factorizations. The **prime factorization** of a number is a string of factors made up only of primes. Below are the prime factorizations of 24 and 60.

$$24 = 2 \times 2 \times 2 \times 3 \qquad 60 = 2 \times 2 \times 3 \times 5$$

Heidi claims that the greatest common factor of two numbers is the product of the longest string of prime factors that the numbers have in common. For example, the longest string of factors that 24 and 60 have in common is $2 \times 2 \times 3$.

$$24 = 2 \times \underline{2 \times 2 \times 3} \qquad 60 = \underline{2 \times 2 \times 3} \times 5$$

According to Heidi's method, the greatest common factor of 24 and 60 is $2 \times 2 \times 3$, or 12.

Heidi claims that the least common multiple of two numbers is the product of the shortest string that contains the prime factorizations of both numbers. The shortest string that contains the prime factorizations of 24 *and* 60 is $2 \times 2 \times 2 \times 3 \times 5$.

Contains the prime factorization of 24
$\underline{2 \times 2 \times 2 \times 3} \times 5$

Contains the prime factorization of 60
$2 \times \underline{2 \times 2 \times 3 \times 5}$

According to Heidi's method, the least common multiple of 24 and 60 is $2 \times 2 \times 2 \times 3 \times 5$, or 120.

Problem 5.3

A. Try using Heidi's methods to find the greatest common factor and least common multiple of 48 and 72 and of 30 and 54.

B. Are Heidi's methods correct? Explain your thinking. If you think Heidi is wrong, revise her methods so they are correct.

■ Problem 5.3 Follow-Up

1. The greatest common factor of 25 and 12 is 1. Find two other pairs of numbers with a greatest common factor of 1. Such pairs of numbers are said to be **relatively prime.**

2. The least common multiple of 6 and 5 is 30. Find two other pairs of numbers for which the least common multiple is the product of the numbers.

3. Find two pairs of numbers for which the least common multiple is smaller than the product of the two numbers. For example, the product of 6 and 8 is 48; the least common multiple is 24.

4. How you can tell from the prime factorization whether the least common multiple of two numbers is the product of the two numbers or is less than the product of the two numbers? Explain your thinking.

Applications • Connections • Extensions

As you work on these ACE questions, use your calculator whenever you need it.

Applications

In 1–6, find the prime factorization of each number.

1. 36   **2.** 180   **3.** 525

4. 165   **5.** 293   **6.** 840

7. Rewrite the prime factorizations you found in problems 1–6 using the shortcut notation described on page 49.

To solve a multiplication maze, you must find a path of numbers from the entrance to the exit so that the product of the numbers in the path equals the puzzle number. No diagonal moves are allowed. Below is the solution of a multiplication maze with puzzle number 840.

Multiplication Maze 840

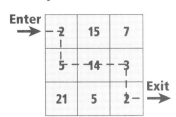

In 8 and 9, solve the multiplication maze. Hint: It may help to find the prime factorization of the puzzle number.

8.   **Multiplication Maze 840**

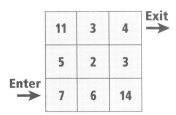

9.

Multiplication Maze 360

2	11	7
15	5	6
3	5	8

Enter → (left of bottom row) Exit → (right of middle row)

10. Make a multiplication maze with puzzle number 720. Be sure to record your solution.

11. Find all the numbers less than 100 that have only 2s and 5s in their prime factorization. What do your notice about these numbers?

12. Find all the numbers less than 100 that are the product of exactly three different prime numbers.

In 13–15, find the greatest common factor and least common multiple for each pair of numbers.

13. 36 and 45   **14.** 30 and 75   **15.** 78 and 104

Connections

16. The number 1 is not prime. Why do you think mathematicians decided not to call 1 a prime number?

Investigation 5: Factorizations

17. a. Find the multiples of 9 that are less than 100.

b. Find the multiples of 21 that are less than 100.

c. Find the common multiples of 9 and 21 that are less than 100.

d. What would the next common multiple of 9 and 21 be?

18. In a and b, use the year you or one of your family members was born as your number.

a. Find the prime factorization of your number.

b. Write a paragraph describing your number to a friend, giving your friend as much information as you can about the number. Here are some things to include: Is the number square, prime, even, or odd? How many factors does it have? Is it a multiple of some other number?

19. Rosa claims the longest string of factors for 30 is $2 \times 3 \times 5$. Lon claims there is a longer string, $1 \times 2 \times 1 \times 3 \times 1 \times 5$. Who is correct? Why?

20. Hiroshi and Sharlina work on weekends and holidays doing odd jobs at the grocery store. They are paid by the day, not the hour. They each earn the same whole number of dollars per day. Last month Hiroshi earned $184 and Sharlina earned $207. How many days did each person work? What is their daily pay?

21. What is my number?
Clue 1 My number is a multiple of 2 and 7.
Clue 2 My number is less than 100 but larger than 50.
Clue 3 My number is the product of three different primes.

22. What is my number?
Clue 1 My number is a perfect square.
Clue 2 The only prime number in its prime factorization is 2.
Clue 3 My number is a factor of 32.
Clue 4 The sum of its digits is odd.

Extensions

23. Every fourth year is divided into 366 days; these years are called *leap years*. All other years are divided into 365 days. A week has 7 days.

  a. How many weeks are in a year?

  b. January 1, 1992, fell on a Wednesday. On what dates did the next three Wednesdays of 1992 occur?

  c. The year 1992 was a leap year; it had 366 days. What day of the week was January 1, 1993?

  d. What is the pattern, over several years, for the days on which your birthday will fall?

Did you know?

If you were born on any day other than February 29, leap day, it takes at least 5 years for your birthday to come around to the same day of the week. It follows a pattern of 5 years, then 6 years, then 11 years, then 6 years (or some variation of that pattern) to fall on the same day of the week. If you were born on February 29, it takes 28 years for your birthday to fall on the same day of the week!

24. Mr. Barkley has a box of books. He says the number of books in the box is divisible by 2, 3, 4, 5, and 6. How many books could be in the box? Add a clue so that there is only one possible solution.

Investigation 5: Factorizations 55

Did you know?

In all of mathematics there are a few relationships that are so basic that they are called "fundamental theorems." There is the "Fundamental Theorem of Calculus," the "Fundamental Theorem of Algebra," and you have found the "Fundamental Theorem of Arithmetic." The Fundamental Theorem of Arithmetic guarantees that every whole number has exactly one longest string of primes, or prime factorization (except for the order in which the factors are written).

Mathematical Reflections

In this investigation, you found strings of factors for a number in the Product Puzzle. You learned to make a factor tree to find the prime factorization for a number. You also learned that the prime factorization of a number is the longest string of factors for the number (not including 1 as a factor). These questions will help you summarize what you have learned:

1 Why is finding the prime factorization of a number useful?

2 Describe how you would find the prime factorization of 125.

3 How can you use the prime factorization of two numbers to determine whether they are relatively prime?

4 How can you use the prime factorization of two numbers to find their common multiples?

Think about your answers to these questions, discuss your ideas with other students and your teacher, and then write a summary of your findings in your journal.

Don't forget your special number! What is its prime factorization?

INVESTIGATION 6

The Locker Problem

You have learned a lot about whole numbers in the first five investigations. In this investigation, you will use what you have learned to solve the Locker Problem. As you explore the problem, look for interesting number patterns.

6.1 Unraveling the Locker Problem

There are 1000 lockers in the long hall of Westfalls High. In preparation for the beginning of school, the janitor cleans the lockers and paints fresh numbers on the locker doors. The lockers are numbered from 1 to 1000. When the 1000 Westfalls High students arrive from summer vacation, they decide to celebrate the beginning of school by working off some energy.

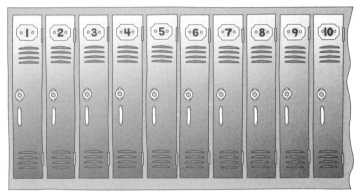

The first student, student 1, runs down the row of lockers and opens every door.

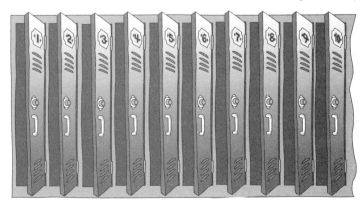

Student 2 closes the doors of lockers 2, 4, 6, 8, and so on to the end of the line.

Student 3 changes the state of the doors of lockers 3, 6, 9, 12, and so on to the end of the line. (The student opens the door if it is closed and closes the door if it is open.)

Student 4 changes the state of the doors of lockers 4, 8, 12, 16, and so on.

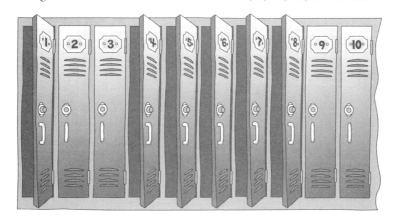

Student 5 changes the state of every fifth door, student 6 changes the state of every sixth door, and so on until all 1000 students have had a turn.

Problem 6.1

When the students are finished, which locker doors are open?

■ Problem 6.1 Follow-Up

1. Work through the problem for the first 50 students. What patterns do you see as the students put their plan into action?
2. Give the numbers of several lockers that were touched by exactly two students.
3. Give the numbers of several lockers that were touched by exactly three students.
4. Give the numbers of several lockers that were touched by exactly four students.
5. Which was the first locker touched by both student 6 and student 8?
6. Which of the students touched both locker 24 and locker 36?
7. Which students touched both locker 100 and locker 120?
8. Which was the first locker touched by both student 100 and student 120?

Applications • Connections • Extensions

As you work on these ACE questions, use your calculator whenever you need it.

Applications

1. What is the first prime number greater than 50?

2. Ivan said that if a number ends in 0, both 2 and 5 are factors of the number. Is he correct? Why?

3. Prime numbers that differ by 2, such as 3 and 5, are called *twin primes*. Find five pairs of twin primes that are greater than 10.

4. What is my number?
   Clue 1 My number is a multiple of 5 and is less than 50.
   Clue 2 My number is between a pair of twin primes.
   Clue 3 My number has exactly 4 factors.

5. What is my number?
   Clue 1 My number is a multiple of 5, but it does not end in 5.
   Clue 2 The prime factorization of my number is a string of three numbers.
   Clue 3 Two of the numbers in the prime factorization are the same.
   Clue 4 My number is bigger than the seventh square number.

6. Now it's your turn! Make up a set of clues for a mystery number. You might want to use your special number as the mystery number. Include as many ideas from this unit as you can. Try out your mystery number on a classmate.

7. **a.** Find all the numbers between 1 and 1000 that have 2 as their only prime factor.

   b. What is the next number after 1000 that has 2 as its only prime factor?

8. The numbers 2 and 3 are prime, consecutive numbers. Are there other such pairs of *adjacent primes?* Why or why not?

Investigation 6: The Locker Problem

Connections

In 9 and 10, describe the numbers that have both of the given numbers as factors.

9. 2 and 3

10. 3 and 5

11. If you find the factors of a number by starting with 1 and finding every factor pair, you will eventually find that the factors start to repeat. For example, if you used this method to find the factors of 12, you would find that, after checking 3, you get no new factors.

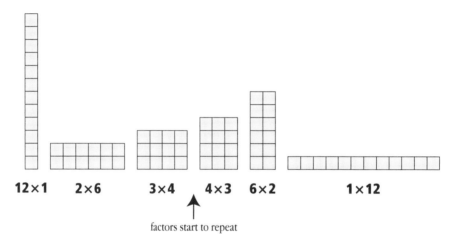

factors start to repeat

For a given number, how do you determine the largest number you need to check to make sure you have found all the factors?

Hint: It may help to first determine the answer for one or two small numbers. For example, you could look at 12 and 16. How would you know, without checking every number, that you will find no new factor pairs for 12 after checking 3? How would you know, without checking every number, that you will find no new factor pairs for 16 after checking 4? It may help to look at the rectangles you made for these numbers.

12. Which group of numbers—evens or odds—contains more prime numbers? Why?

13. Based on what you found out in the Locker Problem, make a conjecture about the number of factors for square numbers. Test this conjecture on all of the square numbers from 1 to 1000.

14. Goldbach's Conjecture is a famous conjecture that has never been proven true or false. The conjecture states that every even number, except 2, can be written as the sum of two prime numbers. For example, 16 can be written as 5 + 11, which are both prime numbers.

   a. Write the first six even numbers larger than 2 as the sum of two prime numbers.

   b. Write 100 as the sum of two primes.

   c. The number 2 is a prime number. Can an even number larger than 4 be written as the sum of two prime numbers if you use 2 as one of the primes? Why or why not?

Extensions

15. Can you find a number less than 200 that is divisible by four different prime numbers? Why or why not?

16. In question 3, you listed five pairs of twin primes. Starting with the twin primes 5 and 7, look carefully at the numbers between twin primes. What do they have in common? Why?

17. Adrianne had trouble finding all the factors of a number. If the number was small enough, such as 8, she had no problem. But with a larger number, such as 120, she was never sure she had found all the factors. Albert told Adrianne that he had discovered a method for finding all the factors of a number by using its prime factorization. Try to discover a method for finding all the factors of a number using its prime factorization. Use your method to find all the factors of 36 and 480.

18. If a number has 2 and 6 as factors, what other numbers must be factors of the number? What is the smallest this number can be? Explain your answers.

19. If a number is a multiple of 12, what other numbers is it a multiple of? Explain your answer.

20. If 10 and 6 are common factors of two numbers, what other factors must the numbers have in common? Explain your answer.

Mathematical Reflections

In this investigation, you solved a problem about open and closed lockers. Then you analyzed relationships among the lockers and the students who touched those lockers. These questions will help you summarize what you have learned:

1 Were lockers with prime numbers open or closed at the end? Explain your answer.

2 Which lockers were open at the end? Why were they open?

3 If factors come in pairs, how can a number have an odd number of factors?

4 Write a problem about students and lockers that can be solved by finding a common multiple.

Think about your answers to these questions, discuss your ideas with other students and your teacher, and then write a summary of your findings in your journal.

Don't forget your special number. What new things can you say about your number?

My Special Number

At the beginning of this unit, you chose a special number and wrote several things about it in your journal. As you worked through the investigations, you used the concepts you learned to write new things about your number.

Now it is time for you to show off your special number. Write a story, compose a poem, create a poster, or find some other way to highlight your number. Your teacher will use your project to determine how well you understand the concepts in this unit, so be sure to include all the things you have learned while working through the investigations. You may want to start by looking back through your journal to find the things you wrote after each investigation. In your project, be sure you use all the vocabulary your teacher has asked you to record in your journal for *Prime Time*.